Sea Bird

Explore the Charming Oceanside Song of Our Shoreline Feathered Friends

Angela Harrison Vinet
& Janis Hatten Harrison

EPIC INK

First published in 2025 by Epic Ink, an imprint of The Quarto Group,
142 West 36th Street, 4th Floor, New York, NY 10018, USA
(212) 779-4972 • www.Quarto.com

Epic Ink titles are also available at discount for retail, wholesale, promotional, and bulk purchase. For details, contact the Special Sales Manager by email at specialsales@quarto.com or by mail at The Quarto Group, Attn: Special Sales Manager, 100 Cummings Center Suite 265D, Beverly, MA 01915 USA.

10 9 8 7 6 5 4 3 2 1

ISBN: 978-0-7603-9357-4

Digital edition published in 2025
eISBN: 978-0-7603-9358-1

Library of Congress Control Number: 2024943784

Group Publisher: Rage Kindelsperger
Creative Director: Laura Drew
Managing Editor: Cara Donaldson
Editor: Katie McGuire
Cover Design: Scott Richardson
Cover Illustration: David Nurney, Diane Pierce, H. Douglas Pratt, and John Sill
Interior Design: Evelin Kasikov

Printed in China

To my mother, Janis Hatten Harrison, whose birding expertise has guided this book. Thank you for sharing your years of knowledge, built up through assorted birding adventures. You're a "hoot."

—*Angela*

Ahoy,
Matey!

Listen up, landlubbers! Along the shores, where water meets land, you'll find one of the most unique—and daring—groups of birds you'll ever see. Those who are patient will overhear the varied songs of these wondrous Sea Birds and witness the gorgeous feathers that make them stand apart from both Country Birds and City Birds. You might even get to see them perform some of their most fascinating feats, diving from cliffs and skimming the seawater for their next meal.

These salty dogs spot fish faster than the captain can say, "Heave, ho," and traveling is no problem, as many of these birds routinely fly to faraway lands over the salty seas. This book is a basic introduction to the seafaring life of birds found along and around the shores of North America.

Mates, unfurl the sails and chase that horizon. You're bound to find the best treasure around—the feathered kind, of course.

Thar She Blows

Spotting Sea Birds Along
Waterways and Shorelines

Travelers best prepare for spotting their feathered friends before a seaside trip. Here are some tips and tricks for spotting a bounty of birds whose sea legs carry them along the shores. Remember, these scallywags are fast and hard to see, so you'd best have a spy scope on hand!

1

Don't forget snacks, water, binoculars, and a field guide. A small collapsible chair to strap to your backpack will also provide needed respite after a hard day's searching.

2

Your birding book or app of choice is the best way identify the treasure you seek.

3

If you're going out adventuring, familiarize yourself with potential birds found at coastal hot spots to make identification a breeze.

4

Bird-watching is best shared with other bird lovers. Check social media for birding groups and see what there is to see.

5

Take heed: seafaring birds are fishermen. Find the food source and you'll spot the bird!

6

Your best bet for spotting Sea Birds is to listen—because those scallywags are loud! Study the calls with free apps and library recordings.

7

Know your feather patterns. Flight feathers are distinctive, so knowing them will make spotting specific Sea Birds from a distance much easier.

8

Know Sea Birds's haunts. Look for them combing beaches, nesting in dunes, walking along rocks, or swarming docks together. Spot one and there's a chance the rest of the crew is nearby.

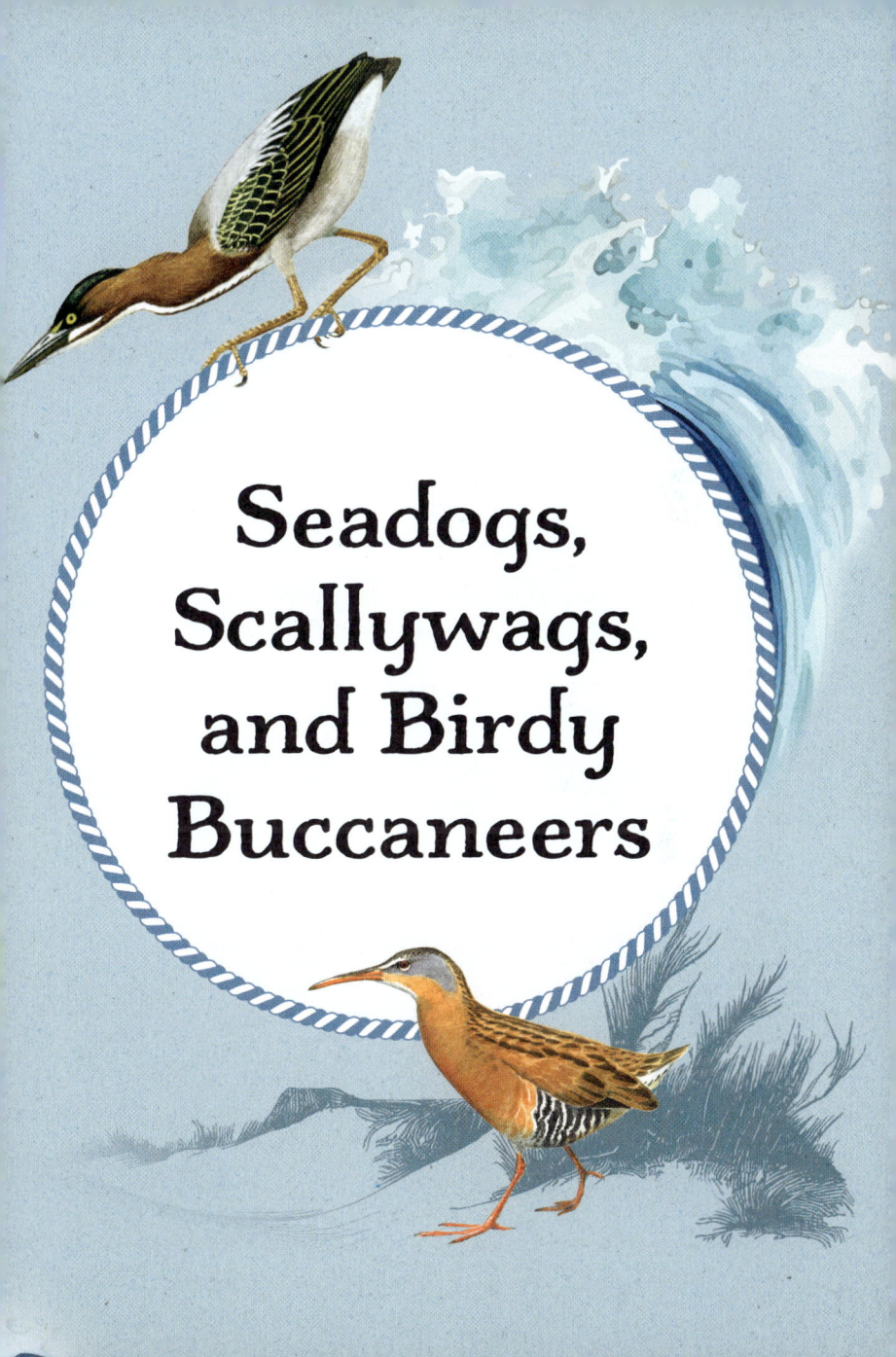

Seadogs, Scallywags, and Birdy Buccaneers

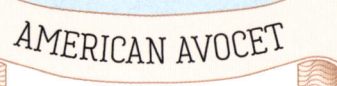

·••· Preening Pillagers ·••·

As delicate as the American Avocet appears, these fierce feathered friends will snatch food faster than a pirate can run a rig. You'll recognize this Avocet by their delicately curved bill and long, slender, blueish legs. This Sea Bird will rest on just one leg when not actively hunting. Typically a coastal bird found in the southern states, this pillaging plunderer is ruthless on the mudflats. Avocets are smart as a tack, fooling predators with fake calls so they don't become someone's next meal. They'll also drop eggs in their friends' nests for safekeeping. As for romance, brooding is for other birds—the American Avocet is too busy scanning the tides for food to be bothered with an elaborate nest. They simply drop some eggs on the ground in the open sun. A mere 24 hours after hatching, the babies get the boot, and they're ready to fend for themselves.

···• Swashbuckling Sword Fighters •···

Standing motionless like a yellow-eyed statue, the American Bittern hides in plain sight. This marsh bird stands at the water's edge, blending into the sea of reeds as they stalk their watery prey. Bugs and spiders stand no chance against these solitary hunters. You'll definitely hear an American Bittern before seeing one. Their rhythmic call rolls over the water, sounding almost like someone beating a hollow tree trunk. The American Bittern's booming calls can be heard all around wetlands and marshes in North America during the nonbreeding season. Territorial males cause plenty of drama within the ranks, sword-fighting like proper buccaneers. These drama kings will bite and shake their prey, and then, after all the chaos, down the gullet goes the meal—headfirst.

AMERICAN COOT

···• Going Overboard •···

What walks like a Duck, talks like a Duck, but isn't a Duck? The American Coot—thanks to their lack of webbed feet. Without those, they have to kick awkwardly to get anywhere by water. Not only clumsy swimmers, these Sea Birds are also awkward flyers. Running on the water, the American Coot takes off at a bit of a tilt, getting themselves airborne, if a bit off-kilter. The American Coot is a common sight in open water, enjoying the company of other waterfowl. Eating their weight in aquatic vegetation and smaller animals, this diving bird will disappear in a flash and pop back up like a buoy. While the American Coot isn't picky about their home, they do look for spaces with plenty of runway for takeoff. American Coots lay their eggs in other Sea Birds' nests, steal food, and consort with all walks of bird life, making them small—but social—bird pirates.

AMERICAN WHITE PELICAN

⋯ Watch Your Booty ⋯

It's a fine day on land or sea when the American White Pelican blows in for a landing—though you'd best watch your catch when they're around. They take what they want, when they want it, without care or concern for their kin. It's a wonder they're able to keep their crisp white feathers pristine, considering how much fishing they do. Scooping up their pay day in those large bill pouches, White Pelicans will give any fishing boat a run for its money. Young White Pelicans can swim with the best of them by the time they're about three weeks old. Considering their nursery is a shallow hole in the ground, this is a handy skill, in case they find themselves swept out to sea when the tides roll in too high. Stealing sticks for their nests and taking fish when they want, the American White Pelican plunders at their pleasure.

AMERICAN WIGEON

···• Green Eye Patches •···

The American Wigeon is a peach of a Sea Bird,
with gentle, soft calls—for a Duck. The males sport
green eye patches and a white head stripe. But that
tough-as-nails exterior doesn't quite match their
personality—this Duck is actually quite the softy.
They'll scatter quickly when disturbed, because no
American Wigeon is looking to make a stand. Often
found in still waters, the American Wigeon sifts
through the top of the water with its wide bill, in search
of plants and insects to make a meal. To impress
their mates, American Wigeons show off some
impressive tail wagging neck thrusts to reel in the
ladies. This Sea Bird nests on dry ground, and
once Wigeons breed, they'll ditch the seagoing life
to become flightless and stick close to home.
The American Wigeon can be found in groups
of their own kind or around other Ducks.

ANHINGA

···• Sea Monster •···

Here, there be monsters, and you'll spot them in both sea and air: keep a sharp eye out for the Anhinga. Looking more like a snake than a bird as it slithers through the water, the Anhinga can fool any lad or lassie new to life at sea. No anchor weighs this bird down, with their light bones and permeable feathers, so they can move through water faster than a crab scuttles across the sand. Once this bird is water-bound, fish had better take heed—the Anhinga has a saber for a beak, ready to spear fish for supper. After a dip in the water, these scallywags proudly dry their feathers, wings outstretched to bask in the glorious sun. While these Sea Birds can swim, nest building is *not* their forte. Their homes are nothing more than a shack of sticks. But why bother building a fine home on land when the sea calls?

BLACK-BELLIED PLOVER

···→ Anchors Aweigh ←···

The Black-bellied Plover casts their cares to the wind when they travel abroad. Located along many areas of the coastal United States—and the world, for that matter—these sailors get around. These adorable little birds are often seen scampering for scraps along the beach, their legs going a mile a minute as they search the receding waves for goodies to gobble. This lot of sailors spend their lives on the ground, often roosting together in the mangroves for protection. Sure to sound the alarm for all Sea Birds, the Black-bellied Plover is a right first mate. Other birds know to take cover when the Plovers signal danger. Plovers have favorite foraging sites among the beaches, mudflats, salt flats, and tidal exchanges they love, returning time and time again. Their large eyes come in handy, allowing them to hunt muddy bottoms at all times, even by moonlight.

BLACK-LEGGED KITTIWAKE

···· Lost at Sea ····

You've never experienced a party like a Kittiwake colony party, not even amongst a pack of sailors on leave. Found all over the world, the Black-legged Kittiwake spends its time chasing prey in the surf. These lifelong seafarers spend their lives on the open ocean, only coming to land to breed and nest. Even when they're nesting, they'll roam nigh on 30 miles out to sea, foraging fish or jellyfish. Despite their delicate features and winning personality, no Sea Bird wants to tangle with this small Gull. Kittiwakes will grab each other's bills and give them a hard twist to make it clear who's boss. One last point of interest: you can always tell a Black-legged Kittiwake by their toes. They have just three, instead of the usual four, with just a nub where the fourth should be.

BLACK-NECKED STILT

···• Crossed Bills •···

Those legs were made for walking, though not along a plank. The Black-necked Stilt is precariously perched atop rosy-red sticks, the longest set of legs this side of a square-rigged schooner. Happily chirping their way through the mudflats, these Sea Birds congregate at low tide or at any muddy mess of a coastline or marsh. The Stilt's excellent eyesight means they can spot even the smallest of prey in the wetlands; larval mosquitos and slimy snails stand no chance. Excellent mates for the marsh foragers, the Black-necked Stilts are the first to send out a distress call to warn the group of danger. These Sea Birds enjoy preening their feathers, digging in the dirt, and gobbling up small aquatic invertebrates. Be warned: the Black-necked Stilt is not to be trifled with, often becoming aggressive toward chicks that roam too close. Still, these Sea Birds are attentive and loving toward their mates, crossing bills to show affection.

···•· The Black Pearl ·•···

Found all along the Pacific coast, the Black Oystercatcher dines on oysters (of course), mussels, and limpets. Their bright orange bill pokes and prods tidal exchanges for treats. These birds strike fast and hard to get the mussels out of their shell casing, hammering at the mollusks with their sturdy beaks. When you stroll the rocky shores and jetties of the Pacific region, you'll see that these Sea Birds stand out in their all-black attire. If you can get a closer look (via spyglass—or binoculars!), take a gander at their black toenails, which help them grip the rocks. Black Oystercatchers can be found along sandy beaches from Alaska to Baja California, wherever there is gravel for their nests. Black Oystercatcher pairs stay together for a season, mostly enjoying the company of just their mate. Having too many birds around is not their cup of tea.

BLACK SKIMMER

⋯• Skimming the High Seas •⋯

The Black Skimmer rides at dawn, sailing through the air at the water's edge. Their black bodies and flaming orange beaks make them regal beauts and signal to others that they don't mess around. A talented hunter, the Black Skimmer uses its beak to skim waters while in flight until they feel a fish with their plow-like beak. Whether the prey is port or starboard, Black Skimmers have a special talent for capturing their meals, thanks to that beak of theirs. Living in colonies on sandy shores, the Black Skimmer loves a good beach or barrier island upon which to settle with their kin. Skimmer chicks are speckled little sand nuggets when freshly hatched, sporting a beak as long as their body (which they'll eventually grow into). While the babies are growing, their parents teach them to loot the beach at the breaking waves, so the young learn at an early age how to successfully pillage and plunder the shores.

BUFFLEHEAD

···• Tiny but Mighty •···

Avast, you Duck lovers, and keep a sharp eye out
for the Bufflehead, America's smallest diving Duck.
The Bufflehead be found in coves along both the
Atlantic and Pacific coasts, with males and females
complementing each other's opposite coloring patterns.
Males have a white head with dark spot feathers and
a white body, while females have a dark head with
white spot feathers and a dark body. Buoyant, beautiful
fluffballs, they're hard to miss floating around bays,
estuaries, and lakes near the coast. Here one minute
and gone the next, the Bufflehead disappears into the
water's depths, catching and swallowing their meal
underwater and eating their weight in invertebrates.
These Sea Birds have been around for quite some
time, as fossils of the species date back hundreds of
thousands of years. Buffleheads love two things: their
mates, who they stay with for many, many years, and a
forgotten woodpecker hole in a tree in which to nest.
A nest and a mate make these Sea Birds happy.

CANADA GOOSE

··· Careeners ···

Despite their name, Canada Geese don't have any particular allegiance to the Great White North. They move around the entire North American continent, though they do breed in Canada. As anyone who's heard them can attest, these Not-So-Sea Birds make a lot of noise, heard far and wide by the distinctive "honk" of their calls. Oft found high in the sky, sailors and landlubbers alike can catch the famous "flying V" formation as the Geese migrate. They stick together and travel in groups, with their young remaining with their parents for at least a year. That's a lot of time together, which explains why Geese can become cranky with each other—though never their mate. If you're new to the birding world, make sure to never call these feathered friends a "Canadian" Goose. Calling them by their proper name just might earn you favor among salty birders.

CASPIAN TERN

···▸ True Landlubbers ◂···

Terns are well known for their regal beach presence, so you're sure to spot the Caspian Tern, the largest of them all. Found globally in bodies of both fresh and salt water, Caspian Terns are a dignified lot, both graceful and deadly—to fish, that is. These Sea Birds frequent most coastal bodies of water and find comfort in being able to see land. No sailor's life for this bunch—they breed in colonies on land. The Caspian Tern's diet consists mainly of fish, but they won't turn down an easy meal of crustaceans or insects. The red bill and black head of the Caspian Tern make it easy to spot this beauty from afar. But do not trespass on their territory—Terns *will* attack to protect their young. Parents feed their young for a long time, wanting to see them reach adulthood before they fly the nest.

···· Casanova of the Sea ····

The Common Goldeneye is a handsome diving Duck. Though their white bodies might get them confused with a Bufflehead, the males' greenish heads and piercing yellow eyes set most straight. They eat mostly aquatic invertebrates and tiny fish they pick up on their diving trips. These strong swimmers are also Casanovas of the sea with dance moves to wow all the females. Common Goldeneyes nest in tree cavities or nesting boxes in northern territories. Ready chicks leap from their perches as Mom carefully watches them tumble out onto the ground—a real leap of faith, with no water to cradle their fall. Though she has her own nest, a female Goldeneye lays eggs in other's nests too, to make sure some of her young survive. While the ladies do squabble quite a bit, defending their territories, the babies sometimes get mixed up with another group during a heated dispute. Birders can hear the Common Goldeneye whistling in flight. They like to sing shanties while they work.

···• Galleon of the Sky •···

With their long, outstretched necks making them look like a Goose in flight, the Common Merganser is an elusive diving Duck. Red-headed females and green-headed males graze in groups, often dipping their heads one after the other while searching for food, like synchronized swimmers doing a well-choreographed routine. Common Mergansers will quickly dive, one after the other, into ice-cold waters to secure fish with their serrated bills. Capable of staying underwater for up to two minutes, these birds glide effortlessly both on top of and under the water. The Common Merganser is easy enough to spot, with its bright red bills and legs, but hard to get close to since they spook easily. These salty dogs especially enjoy perching on rocks or perusing eddies. Baby Mergansers make leaps of faith from their nests in tree cavities or abandoned woodpecker holes to their waiting mothers below. Broods will join together to create a mega-brood, overseen by one mother.

COMMON MURRE

···• Admiral of the Black •···

What do wandering eyes spy when you catch a glimpse of a black and white bird too far north to be a penguin? It must be a Common Murre, all done up in his best formal attire for life at sea. The Common Murre will dive deep into Davey Jones's locker—as deep as 100 feet, give or take—using their wings as fins to fly underwater. These Admirals of the Black only return to land to rear the young in a city of Murres on a cliff. An entire colony of Common Murres living within close quarters be a rowdy lot, but the more the merrier. Noisy calls, squawks, and whirrs mean squabbles, which happen quite a bit. It's hard to get along with your mates all the time, eh? But there is safety in numbers, which far outweighs the disagreements between brethren. Their eggs, more precious than gold and dotted along the cliffside quarters, are teardrop-shaped and bottom heavy, perfect for balancing on the steep cliffsides. Ranging from cream to turquoise, these treasure eggs are a sight to behold.

···• Advanced Anglers •···

With eyes that shine as blue as the sea, the Double-crested Cormorant is easy to spot. This Sea Bird also has not one, but *two* feather tufts behind their eyes to distinguish them from the rest of the smarmy lot. Swimming poses no problem, since their sleek, black wings allow them to cut through the water like the finest sailing vessel you've ever laid eyes upon. After taking a plunge, the Double-crested Cormorant will fan its wings out to dry off in the sun. If you're in need of a sharp-eyed crewman, look no further than the Double-crested Cormorant: they'll out-fish the best of them. And when it comes to more precious bounty, no other seafaring bird loves treasure more. Double-crested Cormorants will plunder the shores in search of special delights and then stash the loot in their nests.

··· Sailing the High Seas ···

The Dunlin is a small bird who likes to eat, drink, and be merry. They arrive in style as a large flock with flags flying, the entire group twisting and turning together in flight like synchronized swimmers. The Dunlin inspects the low-tide mudflats to grab their meals. Estuaries and coastlines make the perfect habitat in which the Dunlin will thrive. They stay out of the waves and focus efforts on the soft beaches and mud, with their beaks always moving as they hunt for food. This bird swabs the deck better than any pirate, cleaning out an area of invertebrates in no time. Though it would be easy to confuse the Dunlin with the Sandpiper, as both birds stay out of the waves, the Dunlin is a shorter and stockier bird. The world-traveling, bug-slurping Dunlins sail the high seas around the world.

···• Looking Good, Mate •···

When birdwatching in the wetlands, if you spy a bird with bright red eyes, chances are it's an Eared Grebe. Looking like a bird with a bad haircut in their nonbreeding plumage, the Eared Grebe still razzles and dazzles with yellow eyelash feathers around their eyes. Eared Grebes look for salty spots to land, gorging on brine shrimp in estuaries. Breeding in colonies, Eared Grebes prefer wet spaces for their nests, often forming a raft for their precious egg cargo. Once the babies hatch, they'll hitch a ride on Mom's back, finding shelter and safety under her wings. Though awkward on land, Eared Grebes are agile in the water, batting those yellow eyelashes at mates and performing a water dance to seal the deal.

···· View from the Crow's Nest ····

With a black head to match their black tipped beak, the Forster's Tern is a distinguished yet delicate bird whose precision with that beak proves deadly for fish. Scanning coastlines and marshes for the shimmering scales of their prey, the Forster's Tern skims the surface of the water to snatch their supper. These birds winter farther north than other Terns and often nest farther inland, too, finding safety in marshes and estuaries. You can distinguish between Terns by looking at their nonbreeding plumage; Forster's Terns have delicate forked tails, black-tipped beaks, and black eyepatches.

Once upon a time, these Sea Birds were listed as a protected species, after the Victorians' taste for their plumage almost put the Terns out of style. Now that the species has rebounded, parents keep a close eye on their babies—the young ones are full of wanderlust.

···• Ruffled Feathers •···

No need for a treasure map to find Gadwalls. These Ducks are found in fresh- and saltwater marshes, donning the finest of feathers. Crisp markings of bold color along with detailed swoops and an almost herringbone pattern in places make it hard to believe the Gadwalls aren't hand-painted by a craftsman. These Ducks, though beautiful, are also tough and will take what they want from other Ducks, often pirating food while others are distracted. During mating season, two besotted Gadwalls will face each other intently and make mirrored head gestures to alert all they are now a couple. Pairs migrate together and honeymoon long enough for the female Gadwall to lay one egg a day, for a total clutch of seven to twelve eggs. Gadwalls are affectionate mates, gently ruffling each other's feathers and communicating through touch. These Ducks have boosted their numbers in recent years thanks to wetland conservation, which makes it easy to find them in the right habitat.

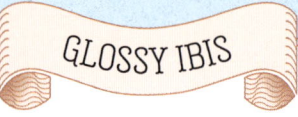

GLOSSY IBIS

⋯• A Mate's Mate •⋯

Rich cinnamon browns, deep greens, shimmering feathers like a sea of gems—if you spot all that and a long, curved bill to go with them, you've got yourself a Glossy Ibis. A well-traveled nomad of a bird who isn't picky about its surroundings, the Glossy Ibis forages with other waterfowl, including Spoonbills, Herons, and Egrets. Any water spot will do, as long as it has plenty of worms and invertebrates for the Glossy Ibis to pillage. When it comes to water, it doesn't matter if it's a flooded field of brackish or fresh water—this elegant beauty will forage and make fast friends. Courting Glossy Ibis bow formally to one another, and then construct bulky nests with sticks in colonies of other Ibis. Glossy Ibis also love to sunbathe, stretching their wings to really show off their stuff. The Glossy Ibis will take to the high seas, crossing the Atlantic Ocean in search of pals across the pond. But, as a true pirate, they love the Caribbean best, haunting the islands year-round.

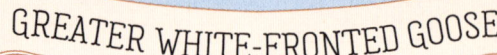

···• Birds of a Feather •···

The saying about "birds of a feather" must've been about the Greater White-fronted Goose, as they live, travel, eat, and rear their young together. Though their signature "honks" are pure Goose, what sets them apart from other varieties is their darker body coloring and bright orange legs and feet. When they're in flight, you can see that the Greater White-fronted Goose has a white band just before their bill, telling birders who's who amongst the Goose population. The Greater White-fronted Goose will winter in large flocks in the wetlands, marshlands, lakes, and rivers close to coastal waters, making the trek down south from the Arctic. Seeds, grains, and fruits are the sustenance they need on the journey down, grazing as they go, while worms and invertebrates are enjoyed once waterside. The families of these Geese form strong bonds, with their young hanging around longer than most birds.

⋯• Warning Shot •⋯

It would be hard to find a Greater Yellowlegs if they stood still, since they're camouflaged perfectly for life in the marsh, but lucky for us, these Sea Birds are constantly on the move. Using those long, yellow legs to wade through marshes, wetlands, and mudflats, they easily disarm small fish and aquatic prey with their sharp beaks. The Greater Yellowlegs needs no compass to crisscross the continent between breeding and nonbreeding times. Natural navigators, the birds know how to get around on pure instinct. The Yellowlegs are anything *but* yellow, keeping a sharp eye out for danger and giving warning calls when needed from the top of their nesting tree. These birds are quite agile and aren't afraid to "bark" in your face, giving their chirpy, back-off sound when you get too close. The Greater Yellowlegs are not to be confused with their kin, the Lesser Yellowlegs. For one, they're . . . well, *greater*, with longer legs and a longer beak.

GREEN HERON

···• Dangerous Beauty •···

Enjoying a sailor's life of sun and fun, the Green Heron haunts spaces from the Gulf of Mexico to the top of South America. Green Herons silently stalk their prey, no matter the time, day or night—they hunt at all times. Though common in the mangroves, the Green Heron has turned up in *uncommon* places after being blown off course and wandering too far. They'll cross the Atlantic and even sometimes find themselves in Europe! These shy birds use tools to lure their prey, often fishing with feathers or bugs to capture larger fish. Their velvet green coloring is striking, but it's a dangerous beauty—this bird is deadly accurate with its bill. They'll mostly perch low to the water to catch prey, but taking a dip for a meal doesn't bother them one bit either. When not fishing, they're concealed in the vegetation near the water as a silent stalker.

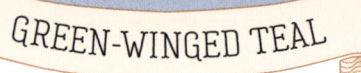

GREEN-WINGED TEAL

···» Any Port in a Storm «···

The Green-winged Teal wastes no time in chick-rearing: within twenty-four hours of hatching, their babies are ready to swim, dive, and forage with the best of them. Easy to spot in shallow ponds or flooded fields, the Green-winged Teal rides low in the water and appears to have no neck when watching from afar. The male's emerald-green hooded head and matching wing patches flash jeweled brilliance just like jewels in the coffer. The rest of their feathers are a delicate herringbone pattern of browns and greens. Chirping their way through the wetlands and marshes, wintering flocks of this Sea Bird can be as big as a thousand birds. For Teals, water is water and they don't care about returning to familiar places and faces—they'll forage anywhere from rivers to puddles. Though the smallest of the Ducks, size matters not to the Green-winged Teal, diving with the best of them for their dinner. Scan the grasses and cattails of marshy wetlands to spy this bird on land.

··· Master and Commander ···

The King Eider is truly king of the water—majestic in every way, from their crown of back feathers to their unmistakable head of blues and grays, with a pronounced, bulbous beak. Nesting in the tundra and talking shop with their fellow seagoers around the pack ice, these Sea Birds eat shellfish high in the Arctic. King Eiders are masters at underwater diving, using their wings to fly through the water at depths of 80 feet. Your best chance to spot the King Eider is to venture north, where the wind bites and the seas hold dangerous frozen icebergs. If you aren't built for life at sea, search for these Ducks where they nest on rocky ridges in the salt- or freshwater tundra, concealed by vegetation. You won't find a tougher bird—a King Eider mother will sacrifice herself for her babies. When threatened, she digs in and flattens herself onto her eggs to keep them safe. Once the chicks are ready, the flock heads to the ocean to begin life at sea.

⋯• Sea of Thieves •⋯

Take heed: *never* trust a Laughing Gull. These grubby mites will rip food from your hand (or your beak) before you've had a chance to take a bite. Most equate the calls of this Gull with summer and sun, but the Laughing Gull is more than a symbol of surf and sand—these birds are a lifestyle. Laughing Gulls are easy enough to spot on a day at the beach, with their dark head feathers paired with a soft gray/white body. What they lack in looks, they make up for in cunning. Lonely males set to building a nice ground house, trying to rope in a female to be his first mate. Once paired, they'll toss eggshells overboard after the fluffy, sandy chicks hatch. Beggars can't be choosers, so the Laughing Gull will digest just about anything, from berries and bugs to other bird chicks and maybe their eggs too.

⋯• Landlubber's Life for Me •⋯

It's a known fact that Petrels love a life at sea. But one variety of this Sea Bird is more of a landlubber than the others: the Leach's Storm-petrel. Though they love the sea air, they'd rather find themselves marooned on a beautiful deserted isle—especially when it comes to courting. They'll chat up their mates—of the loving kind, not the ship-running kind—into the wee hours of the night. With their forked tail and white rump, you'll need a pretty fast boat to catch sight of this Sea Bird while the sun is blazing. To find the Leach's Storm-petrel's nesting spots, climb into the crow's nest and look out for islands far off the coast.

LONG-BILLED DOWITCHER

···· A Right Jolly Lot ····

Long-billed Dowitchers eat together, fly together, and abandon ship together. Wading through murky waters, the Long-billed Dowitcher rapidly pokes and prods for food along the way. What's truly remarkable about this bird is its sixth sense for finding food, using receptors in their bill as they probe the mud for food. Within the Dowitcher family, there are just two types—the long-billed and short-billed—which makes it easy on an old sailor's eyes to tell them apart. This Sea Bird is seasonal, weighing anchor in the brooding season to travel to the northern tundra in search of a muddy, sandy floor to raise their young. Nesting in a specially made cup of grasses, Dad rears the young once they hatch. The best places to spy the Long-billed Dowitcher in their nonbreeding season are tidal exchanges, mudflats, and marshes with shallow wading water.

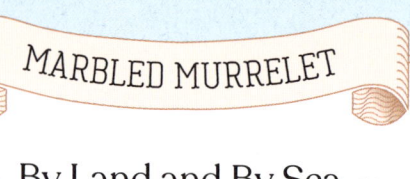

MARBLED MURRELET

···▸ By Land and By Sea ◂···

The Marbled Murrelet is a Pacific Coast diving, flying Sea Bird whose life at sea is only deterred by their desire to raise a family. Don't let the Marbled Murrelet's drab attire fool you—this is one special bird, as they change colors with the breeding seasons, from browns to grays. Murrelets enjoy the finer things in life, like old trees, which they'll scout for nesting. They don't want any old nest, but a special, moss-covered tree limb aged at about two hundred years. After courting at sea, Marbeled Murrelet pairs lay their eggs in the natural crooks of the old tree, using lichens and velvety green mosses as their nursery. Mated pairs will playfully chase each other in the water, swimming side-by-side before returning with fish for the babies. These Sea Birds fish by day along river mouths and tidal rips and parent their young in the forest at night.

···❖ Dead Wrens Tell No Tales ❖···

Perched on stalks of grass or cattails, the Marsh Wren sings shanties to signal friends and potential mates alike. You'll hear them first, then you can search the grasses for a small bird with an upturned tail. The curved bill of the Marsh Wren is important for grabbing spiders from their webs and other insects from their habitat. This practice helps the birds gain energy for more amorous adventures. The males are ambitious, building 10 to 20 nests in their territory in hopes of finding more than one bonnie lass with whom to raise young. Flitting and fluttering about, these birds are quick and nimble, but don't be fooled—Marsh Wrens are harsh. They completely ravage other Marsh Wrens' nests and eggs, and this practice often bleeds onto other marsh birds as well. No one builds in a Marsh Wren's territory without a fight.

···· **Blow the Man Down** ····

Master of the sails, the Northern Fulmar may look and sound like a Gull, but don't be fooled. Well-traveled sailors can spot the difference between the Northern Fulmar and a Gull by watching their wing beats: Fulmars beat their wings faster than the Gulls. There are two varieties of Fulmar—the light and the dark morph—but be assured that both will poach fish from vessels at sea. While the Northern Fulmar will dive and forage, this Sea Bird loves an easy meal of cast-off fish parts from fishing boats. You'll spy this Fulmar in the north, where they haunt the rocks and cliffs for nesting. There's nothing this savvy lot can't do: they swim, they fly, they shallow dive, and they rock climb. They get a running start to take off from the water, searching high and low for a quick meal. That bump on their nose is more than a beauty mark— Fulmars can shoot salt right out of that nose blowhole. And if you get too close, they'll spew putrid fish guts back at you. Best to watch the Northern Fulmar from afar.

·⋯· Even Keel ·⋯·

Spotting an upturned rump with long tail feathers on the horizon? It's probably a Northern Pintail. The quintessential "quack" of these Ducks can be heard rolling over the water for quite a distance. Full of elegance and grace, these Ducks glide easily over the water and will go head over heels to find their favorite foods underwater. Swimming around wetlands, bays, and lakes around the world, the Northern Pintail lives much longer than a seafaring pirate. The northern portion of their namesake comes from their love of nesting in the northern country, as soon as ice on the waterways breaks up. They like to migrate with other Ducks, flying in arranged long, wavy lines in the evenings and under cover of night. These Sea Birds sit higher than the rest of the Ducks, making them easier to spot. Though they nest far from the ocean, they do sometimes stretch their sea legs to travel to the coasts. These social birds are even keel, never causing drama within their groups.

NORTHERN SHOVELER

··· Shoveling for Their Supper ···

The Northern Shoveler lives up to its name, combing the water for tasty invertebrates and seeds. They spend a lot of time with their bills underwater, which is a shame—Northern Shovelers are striking. Their bills are as large as their heads, which are a distinct green in males, matched with yellow eyes and white-striped tail feathers. Northern Shovelers have developed a water dance of sorts, spinning to stir up food from the bottom. Rather awkward on land, Shovelers are graceful in the water, whether it be a wetland, marsh, or estuary. But perhaps the most important thing to note about the Northern Shoveler is to avoid scaring a mother on her nest. She will defecate on her eggs faster than Captain can yell, "Swab the deck," to deter predators from making her babies their next meal.

PACIFIC LOON

··•· Shoving Off ·•··

As their name suggests, Pacific Loons are found on the Pacific coast (and in large bays during nonbreeding season). Gently gliding across waters, smooth as a clipper, a Pacific Loon can dip its head into the water without putting one feather out of place. Their feathers always look more like fine art, with delicate brushstrokes of black on white and white on black, and they are water repellant. Brilliant red eyes help them spot and catch fish as they dive much deeper than their brethren. You'll find Pacific Loons on larger bodies of water, since they need an enormous runway to take off for flight; at bare minimum, they need a fathom. Once in the air, they fly at speeds of over 35 miles per hour. Their love of swimming is evident, spending most of their time in the water, then retreating to nest on tundra lakes.

··· Terror on the High Seas ···

Even savvy birdwatchers might sometimes confuse the Pomarine Jaeger with its counterpart, the Gull, but the Jaeger is a more aggressive bird with different coloring patterns. A sea bully, the Pomarine Jaeger spends most of its time at sea, forcing other Sea Birds to hand over their catch. With pure brute force, the Pomarine Jaeger sacks other birds' meals and plunders as they go, striking fear into the hearts of birds between Alaska and the high Arctic. Small rodents are usually on the menu, but nothing compares to the sweet taste of lemming. You'll have to book a special voyage to find this particular bird.

⋯• No Sitting Duck •⋯

Redheads are so eager to find friends, they'll even land in a group of decoys in search of a good time. This Duck mostly enjoys open-water lakes and coastlines, any place where the entire flock can sample the sea grass. When it's time to nest, female Redheads have no interest in becoming "sittin' ducks." In fact, they let someone else do the dirty work. While Redheads will make a nest of their own, they find great joy in giving others their young to raise, sticking their eggs in all kinds of waterfowl nests. A Redhead will make a nest of her own, but she really loves having others do the child-rearing. When it's time to settle down, the males will do impressive neck bends, touching their tail with the backs of their bills to impress their lovely ladies. Though the females are drab, the males make up for it tenfold with their brilliant, cinnamon-red heads and yellow eyes. In summer months, look for Redheads along reedy ponds, gobbling midges and mayflies. Fish eggs and snails are treats for them too.

RING-BILLED GULL

···· Savvy Scavengers ····

Fast, fearless, and familiar, the Ring-billed Gull is found all over North America. If you spy a Gull far from the coastline, you've probably found a Ring-billed Gull. While they've adapted to life at any watery spot, these birds thrive in and around human environments. They've taken a special liking to life at landfills. Opportunist and savvy, the Ring-billed Gull is a true survivor, not picky when it comes to where they get their next meal. They'll steal and scavenge, like all good pirates do, but perhaps the most endearing thing about the Ring-billed Gull is that they always return home. They return to where they were born and raise their own young in the same nesting grounds. With an appetite for all things food, the Ring-billed Gull lives a long and merry life, filling the air with those all too familiar sea squawks.

···▸ Beachcombing Beauties ◂···

If it's Ruddy Turnstones you're after, it's best to search beaches and shorelines for birds literally turning stones for food. Flipping rocks, debris, and discarded shells comes easy to the Ruddy Turnstone since they've got a sturdy beak perfect for the job. These birds even have a special toenail for gripping the side of rocks while they search for food. Champion beachcombers, these birdies gobble up all the sand fleas and flies, saving beach lovers from pesky biters. These Sea Birds blend in perfectly with the debris line left by the high tide. At mating time, their breeding plumage takes on a calico coloring of browns, blacks, whites, and oranges to match the lichen and rocks of the tundra floor. Ruddy Turnstones are mighty creatures of habit, returning to the same nesting site year after year, and their chicks leave the nest mere hours after hatching.

SORA

··· Castaway by Choice ···

The secretive Sora can be found in the wee hours of the morn or late in the evening, in both fresh and brackish wetlands. Hiding out in the dense grasses, reeds, and cattails until it's time to hunt, the Sora is an illusive Sea Bird and, despite their abundance, they are difficult to find if you're not savvy. You can identify them by their signature short, yellow beaks. But when you can spot one, you'll understand why the Sora's tubby stature has endeared them to birders. They peck at the water in search of aquatic seeds and invertebrates, always ready to zoom back into the safety of the vegetation at a moment's notice. After courting and claiming a mate by staring deep into each other's eye for an awkward fifteen to thirty minutes, Soras will weave grass nests on top of vegetation near the water. You'd be a fool not to love a Sora.

···• Sound the Trumpets •···

At about six feet in length, the Trumpeter Swan is one of North America's largest waterfowl and certainly one of the heaviest flying birds. Known for their grace and snowy-white feathers, the Trumpeter Swan is a sight to behold—especially since they've come back from the edge of extinction after being hunted for their gorgeous feathers. Though mainly northern birds found around the coasts of the Pacific Northwest, some lucky few might spot Swans in northeastern wetlands or around the Great Lakes during their breeding season. These pretties preen as they glide through quiet waters just a few feet deep. The only things they need are plenty of space and vegetation, the former to have room for takeoffs and landings and the latter to feed their appetites. Trumpeter Swans typically mate for life and build floating nests on top of existing beaver or muskrat homes.

TUFTED PUFFIN

···❖ Sea Legs Required ❖···

Tufted Puffins are dazzling Sea Birds whose black body feathers are complemented by a bright orange beak and a glorious mane of golden feathers. Only found in the chilly northern Pacific waters off North America, this showstopper spends most of its life at sea. Tufted Puffins "fly" underwater in a deep dive, sometimes getting as deep as 360 feet, using the currents to catch fish and stuff them sideways into their bills. These clever birds will even use a fish's bioluminescence against them for a little night fishing, following the fish's light right to dinner. Tufted Puffins eat underwater, so if you see one carrying fish in its mouth, it is most certainly for the babies. Courting is something worth watching as well, if you've got your sea legs. Tufted Puffins perform a water ballet to intrigue their beloved, then nest on cliffs around islands in the North Pacific.

VIRGINIA RAIL

··· Skittish Sailors ···

The Virginia Rail is not adventurous, often seen nervously twitching their tail feathers or hiding in the mud of the freshwater marshes they so enjoy. (Lucky for them, their coloring is nearly a perfect match for their surroundings—excellent camouflage.) Virginia Rails push nimbly through dense vegetation when it's time to abandon ship, preferring to scatter on land than take refuge in the water. Cattails and rushes provide the perfect habitat for the Virginia Rail to weave their floating basket palaces. Nesting pairs will do everything possible to make sure their young survive, including making decoy nests. To find this shy bird, check tidal exchanges at dawn and dusk for a solitary bird—they really aren't social until they find their mate for the season.

WHIMBREL

···> Riding Fair Winds <···

It's hard to miss a Whimbrel in flight. A slender, curved bill—perfect for hunting in mud—and sandy brown feathers help this Sea Bird blend in perfectly at the mudflats they frequent. Skillfully slicing into the sand to nab delicious invertebrates, the Whimbrel plunders crab holes with the best of them. This Sea Bird will lower the boom on the crab claws, knocking them off and washing the crab before the whole thing goes down the gullet. Since these Sea Birds have a taste for the finer things, fiddler crabs are often on the menu, though they won't turn their beaks up at some fancy berries. Basically, any food in the intertidal zone is theirs for the taking. Singing as they glide and migrate over the open ocean, the birds cover some ground on fair winds, from the very top of North America all the way to Central America.

⋯•⟩ Night Shift ⟨•⋯

We all know it takes two to tango, but only Daddy Willet sleeps at the nest at night. Mom will sometimes feign injury to lure predators away from the nest. After a good day's work around the nest, she's off in the evenings to pursue other interests. The Willet braves the coasts with their long sea legs as they chase prey in the waves breaking upon the beach (they especially enjoy shellfish and crabs). To correctly identify a Willet, check for the white wing band visible in flight. Willets defend their territory from plunderers—you've been warned. Once considered a tasty dish, Willets are now a protected species and have prospered since earning the designation. Because they make nests near the water in sand dunes or grasses, their babies would fall prey to some other Sea Bird's dinner if not for Dad spending the night. Adult Willets can feed both day and night, since special receptors in their sensitive beaks allow them to detect and locate their favorite aquatic invertebrates.

WILSON'S PHALAROPE

··· Dandy Dancer ···

The Wilson's Phalarope jigs its way through life, all the while pecking bugs out of muddy wading waters. The bird's long, slender beak, as precise as a needle, is perfect for plucking insects from the water. Wilson's Phalaropes twirl in the water to create tiny whirlpools that sweep the bugs off their many feet, so the birds can make quick work of them. In this Sea Bird's world, the females have the brilliant coloring, as they're the ones doing the courting. Once mates are chosen, the males are left to sit on the nest with the eggs, while the female Phalaropes let their wanderlust take them wherever strikes their fancy. Descending upon the salt flats in great numbers, Wilson's Phalaropes will eat and eat and eat until they're so large they can't even fly. These Sea Birds migrate in large flocks in late summer, traveling to any shallow waterhole full of bugs.

WILSON'S SNIPE

⋯ Fool's Errand ⋯

Beware the pirate who sends you on a Snipe hunt—it's a fool's errand. The Wilson's Snipe is an elusive bird, difficult to find even with the best spy glass. If you're lucky enough to spot one, you'll see a bird with an abnormally long beak and a short, squat body. One of the reasons for their stocky appearance is their extra-large pectoral muscles, which also allow them to fly about 60 miles per hour. One might think they had too many long days on the high sea, what with their erratic, zigzagging flight pattern. With a deeper eye socket than most birds, the Wilson's Snipe can see both forward and backward—quite a handy parlor trick. This Sea Bird thrives in low-lying vegetation, where they blend in perfectly with their surroundings thanks to the yellow stripe down their backs—it helps them look just like the reeds laying on the ground. However, there are a few tips for spotting a Snipe. This lot is more active at dawn and dusk, and they're typically found in North American fields and wetlands, searching for their favorite meal: insect larvae.

WOOD DUCK

···• Walk the Plank •···

Wood Ducks are one of the most elaborately feathered Ducks, found in North America year-round. These jewels of the Duck family nest in tree cavities along the water's edge; look for them in wooded swamps around creeks, rivers, and swampy waters. Wood Ducks are masters of repurposing, using natural cracks in the tree or around a branch that's broken away to make their home. Those smaller openings keep the Wood Ducks safe from predators. Wood Ducks love nesting over slow-moving water, so their babies have a safe drop zone after walking the plank. Wood Ducks will have two broods a year, lining the nest with Mom's feathers. Birders can put up wooden duck boxes to entice them to make a home nearby.

Aye, Aye, Captain

No matter the type of sailor they are, whether salty dogs or sunset cruisers, Sea Birds fascinate and entrance onlookers with their plumage, pillaging, and piracy. With personalities all their own, each bird of the sea uses their savvy sea knowledge to survive—and thrive. Some birds have adapted to coexist with humans, seeking them out to steal a meal, while others have moved farther away from human interference.

Finding niches within their environments, each Sea Bird has unique characteristics that set them apart from others. Though they may sometimes squabble, these Sea Birds' adaptations and habits ensure the survival of every feathered sailor. Every bird has their own temperament and personality. All one must do to discover them is sit quietly and observe to discover the pirates amongst us.

Sea Birds are worthy of our admiration—and protection. Most of these birds depend heavily upon the barrier islands or sea dunes to raise their young, feed, or catch the eye of a mate. It is up to us, the humans of their world, to protect the habitats necessary for their survival for generations to come.

Washington

Montana

North Dakota

Oregon

South Dakota

Idaho

Wyoming

California

Nevada

Utah

Nebraska

Colorado

Kansas

Oklahoma

New Mexico

Arizona

Texas

Alaska

1	American Avocet	11	Black Skimmer
2	American Bittern	12	Bufflehead
3	American Coot	13	Canada Goose
4	American White Pelican	14	Caspian Tern
5	American Wigeon	15	Common Goldeneye
6	Anhinga	16	Common Merganser
7	Black-bellied Plover	17	Common Murre
8	Black-legged Kittiwake	18	Double-crested Cormora
9	Black-necked Stilt	19	Dunlin
10	Black Oystercatcher	20	Eared Grebe

X Marks the Spot:
A (Treasure) Map of Sea Birds

Vermont

Maine

New Hampshire

Massachusetts

Rhode Island

Conneticut

New Jersey

Delaware

Maryland

North Carolina

South Carolina

Minnesota

Wisconsin

Michigan

New York

Iowa

Illinois

Indiana

Ohio

Pennsylvania

Missouri

Kentucky

Virginia

Tennessee

Arkansas

Georgia

Alabama

Louisiana

Mississippi

Florida

21	Forster's Tern	31	Long-billed Dowitcher	41	Ruddy Turnstone
22	Gadwall	32	Marbled Murrelet	42	Sora
23	Glossy Ibis	33	Marsh Wren	43	Trumpeter Swan
24	Greater White-fronted Goose	34	Northern Fulmar	44	Tufted Puffin
25	Greater Yellowlegs	35	Northern Pintail	45	Virginia Rail
26	Green Heron	36	Northern Shoveler	46	Whimbrel
27	Green-winged Teal	37	Pacific Loon	47	Willet
28	King Eider	38	Pomarine Jaeger	48	Wilson's Phalarope
29	Laughing Gull	39	Redhead	49	Wilson's Snipe
30	Leach's Storm-petrel	40	Ring-billed Gull	50	Wood Duck

Land, Ho!

Whether they seek their fortunes on land or at sea, North America's feathered friends fly many miles in search of climates that provide the means necessary for their survival—without a treasure map. Often trekking across the continent and the Gulf of Mexico to find a place that can support their nautical lifestyle, these birds naturally know their destinations for food and friends.

During the winter months, food is scarce and difficult to find in cooler climes. By traveling to warmer ports, the birds can easily find their culinary favorites, from the sweet nectar of flowering plants to a bountiful buffet of bugs. Scientists are not exactly sure how the birds know when and where to go, but the birds seem to have an internal compass to guide them to places they've never been. They also seem to know instinctually the perfect time to travel abroad, like a true sailor when the winds change.

Barrier islands just off America's Gulf Coast states serve as more than pretty stretches of beach. These islands provide the perfect place to drop anchor and are a welcome sight to the birds traveling home from South America. Migrating birds arrive in time for spring in the northern hemisphere. People flock (pun intended) to these spaces to catch glimpses of birds not typically found in their area. The birds find safety (and plenty to eat!) within the green space after their exhausting journeys across the gulf.

When conditions are right, a fallout can occur at these first sight land masses. This phenomenon occurs when storms or high winds blow exhausted birds onto land after they've crossed the Gulf of Mexico—they quite literally "fall out" of the sky. Bird enthusiasts and ornithological study groups will take the opportunity to band birds for research purposes, putting a small band with their current coordinates on the bird's leg. Birds can then be tracked and reported when sighted to assist in data collection. Along the gulf states, there are many sites where volunteers can help catch, band, and release birds for ornithologists.

Birds often deplete their stored energy on the long flight, barely making it to land after a crossing. While there are casualties in the Gulf, the vast majority of birds make it, if slightly worse for wear. After their long migration journey, birds will catch some R & R on these first sight land masses before packing their bags and traveling on to the next adventure.

On the Lookout

Once you've climbed down from your crow's nest, make note of the Sea Birds you've spotted on your latest voyage.

Type of Bird

Date

Location

Notes

Type of Bird

Date

Location

Notes

Type of Bird

Date

Location

Notes

Type of Bird

Date

Location

Notes

Type of Bird

Date

Location

Notes

Type of Bird

Date

Location

Notes

Type of Bird

Date

Location

Notes

Type of Bird

Date

Location

Notes

Type of Bird

Date

Location

Notes

Type of Bird

Date

Location

Notes

Type of Bird

Date

Location

Notes

Type of Bird

Date   Location

Notes

Type of Bird

Date   Location

Notes

Type of Bird

Date   Location

Notes

Type of Bird

Date   Location

Notes

Type of Bird

Date

Location

Notes

Type of Bird

Date

Location

Notes

Type of Bird

Date

Location

Notes

Type of Bird

Date

Location

Notes

Meet the Crew

Angela Harrison Vinet and **Janis Hatten Harrison** are a mother/daughter duo who have been writing these bird books during Janis's battle with bone cancer. They have found great joy amid darkness while working together on this series. The perfect pair, Janis is the birding expert and comic relief while Angela is the writer and teacher at heart.

Me Hearties

Thank you to the Cornell Lab's All About Birds website and the National Audubon Society's online Guide to North American Birds. These free, helpful websites allow many to learn about the birds we all love to watch through their user-friendly platforms.

Education is at the heart of good things. To teach someone is to become a part of them and their journey through life, because learning never stops. Two teachers who gave me a chance are Mrs. Donna Underwood and Mrs. Mary Walker. Thank you for believing in me.

Thank you to my sister, my biggest supporter, whose solid mind keeps me from floating in the clouds—I cannot do it without you.

Thank you to my children, Julien, Will, and Christian, who have given up time with me so that I could write these books. You are my first and last thoughts of the day. I love you with my entire being.

—Angela